BARBED WIRE

BARBED WIRE

A POLITICAL HISTORY

Olivier Razac

TRANSLATED FROM THE FRENCH BY
Jonathan Kneight

THE NEW PRESS

NEW YORK

© 2000 by La Fabrique-éditions.

English translation © 2002 by The New Press

LIBRARY OF CONGRESS CATALOGING-IN-PUBLICATION DATA
Razac, Olivier.
[Histoire politique du barbelé. English]
Barbed wire: a political history / Olivier Razac; translated from the French by
Jonathan Kneight.
p. cm.
Includes bibliographical references and index.
ISBN 1-56584-735-0 (hc.)
1. Barbed wire—Political aspects. 2. Wire fencing—West (U.S.)—History.
3. Wire obstacles—History. 4. Concentration camps—History. I. Title.
TS271 .R39 2002
323.4'9—dc21
2002019536

Originally published as *Histoire politique du barbelé: La prairie, la tranchée, le camp*
by La Fabrique-éditions, Paris

Published in the United States by The New Press, New York, 2002

Distributed by W. W. Norton & Company, Inc., New York

The New Press was established in 1990 as a not-for-profit alternative to
the large, commercial publishing houses currently dominating the book
publishing industry. The New Press operates in the public interest rather than
for private gain, and is committed to publishing, in innovative ways, works
of educational, cultural, and community value that are often
deemed insufficiently profitable.

The New Press, 450 West 41st Street, 6th floor, New York, NY 10036
www.thenewpress.com

Printed in the United States of America.

Book design by Lovedog Studio

2 4 6 8 10 9 7 5 3 1

CONTENTS

It's the best fence in the world.
As light as air.
Stronger than whisky. Cheaper than dust.
All in steel and several miles long.
The cattle haven't been born that can get
through it.
Gentlemen, take up the challenge and
  bring your cows.

–John Warne Gates,
barbed-wire merchant,
Texas, 1870s

INTRODUCTION

For over a century, iron barbed wire has been used almost everywhere: around fields and pastures in the country, in the city, on walls or fences of factories, around military barracks and even private residences. It is also found along national frontiers, on battlefields, and for holding people—to protect them, expel them, or kill them.

Yet this object, devised at the lowest level of technology, is not especially sophisticated. In a century of stunning technological advancements, when a computer's power becomes laughable in ten year's time, when obsolete objects pile up in the junkyards of modernity, barbed wire has remained almost unchanged since its inception.

It has always been efficient enough to perform its designated tasks: to define space and to establish territorial boundaries.

The perfection of a tool of power is not measured so much by its technical refinement as by its economic adaptation. The instruments which serve authority best are those which expend the smallest amount of energy possible to produce the effects of control or domination. Barbed wire is such an instrument, because its simplicity makes it a cheap and supple tool, adaptable to all kinds of uses.

Yet merely because barbed wire has always been successful does not mean that it is still the foremost technology for managing space. A certain trend is emerging by which power, with the greatest discretion, is occupying space. But is this something new? When barbed wire first appeared, power was already rejecting the thickness of stones, massive separations, to create territorial divisions. Yet the appearance of barbed wire also foretold its eventual obsolescence, the time when it would be too visible and too heavy and thus would have to be replaced by more ethereal means of controlling space through the use of more furtive instruments. These latter tools establish limits not of wood, stone, or metal but of light, waves, and invisible vibrations.

1

THREE
HISTORICAL
LANDMARKS

Barbed wire is everywhere, and since its invention has been used throughout the world, in all kinds of ways and with different objectives, even contradictory ones. For these reasons, its history seems too chaotic to chart.

This work does not attempt to be a comprehensive history of barbed wire. Rather, it isolates those appearances of barbed wire which convey the clearest and most significant political implications. For this purpose, three of its historical uses will be discussed: the American prairie, the World War I trenches, and the Nazi concentration camps.

In these three cases, we see the use of barbed wire in its pure form. Here, it is not something merely added to other elements, as is the case, for example, when it is stretched on top of a wall. Rather, it constitutes the essential and primary material used to establish specific boundaries.

In these three cases as well, its use goes far beyond its primary agricultural purpose. Barbed wire's direct political impact has been crucial to three disasters: the physical elimination and then the ethnocide of the North American Indians; the unthinkable bloodbath of modern war; and at the center of the totalitarian catastrophe, the concentration camps and the Jewish and Gypsy genocides.

AMERICA:
FENCING IN THE PRAIRIE

The Invention of Barbed Wire and
the Conquest of the West

1874: an obscure date, and yet decisively important in the history of the United States. An Illinois farmer, J. F. Glidden, took out a patent for the barbed iron wire he had invented. He was not the first to propose something like this, but his invention was technologically superior to other new sorts of fences. It consisted of only two iron wires with a series of barbs (pieces of twisted iron whose two ends are beveled and thus sharpened). At first, Glidden limited himself to twisting each barb on a single wire, but the barbs' grip loosened quickly and they moved up and down the wire. Then Glidden had the clever idea of reinforcing his tool by rolling a second wire around the first and its barbs. Now the barbs could be kept in place, and the whole structure was much more resistant.[1]

In fact, four-fifths of the new ideas for fences proposed in the mid-nineteenth century came from the Middle West. Farmers were beginning to invade the Great Plains west of the Mississippi and needed some cheap means of fencing off their fields. Yet at about the same time, starting with the 1850s, after decades of regular colonization, the advance toward the West was slowed down, mainly by the geographical conditions of the territory beyond the 100th meridian, which cuts the United States in two. The lands between the Missouri River and the Rocky Mountains were reputed to be uncultivatable; they were arid and lacked wood and stone. This "Great American Desert" was still largely unexplored by 1850. Perhaps only the metaphor of the ocean can evoke this landscape of gentle reliefs whose surface is covered by high grass perpetually agitated by a movement which resembles waves; a smooth space over which glided the caravans of the first pioneers, but, long before them, the herds of bison and the Indians who hunted them. The hostility of

A barbed-wire fence encloses "the Great American Desert."
Day County, South Dakota. John Vachon, photographer, November
1940. Farm Security Administration—Office of War Information
Photograph Collection. (*Library of Congress*)

the land to agriculture and therefore to settlement would not keep out the white man much longer.

The 1862 Homestead Act expressed the government's will to accelerate the westward expansion under pressure from poor, landless farmers. The act gave any American citizen free ownership of 160 acres of public land, on the condition that it be cultivated. Yet, rather than benefiting the poor farmers, the act enriched private speculators, in particular the railroad companies. The Homestead Act thus marked the last stage in the American colonization of the West. Still, cultivating the desolate and inhospitable western lands remained problematic.

As early as 1865, big ranchers began exploiting the flat Texan plains, grazing immense herds of longhorns and driving them to the railheads north of Texas (Abilene, Cheyenne, and so on). From there, the cattle made the cheap and easy trip east (to, for example, Kansas City and Saint Louis), where their price increased tenfold. For twenty years, cowboys drove millions of cattle through territory formerly inhabited by bison and Indians, making the fortunes of a few cattle barons. Quite soon, however, it became clear that open grazing was neither practical nor feasible. The herds were excessively exposed to the extremes of climate and to the dangers of the drive

Putting up a barbed-wire fence on the Milton farm
at El Indio, Texas. Russell Lee, photographer,
March 1939. Farm Security Administration—Office
of War Information Photograph Collection. (*Library
of Congress*)

Barbed-wire fence poles. Stark Country, North Dakota.
John Vachon, photographer, February 1942. Farm Security
Administration—Office of War Information Photograph
Collection. (*Library of Congress*)

north. Furthermore, as early as the end of the 1870s, farmers who enclosed their fields reduced the number of cattle trails, provoking veritable "barbed-wire wars." "The empire of cattle collapse[d] as rapidly as it was born. And if cattle raising continue[d] on the Great Plains, it [was] done in a much less adventurous and picturesque way."[2] However, that short period of the open range, of herds and of wandering cowboys, all destined to become a central myth of American history, led to exploring the great prairies, rendering them less threatening and more accessible.

Finally, several technological innovations enabled the farmers to settle on and cultivate the great prairie. The transcontinental railroad, inaugurated in 1869, had immense consequences in joining the lands of the West with the great economic and industrial centers of the East, and in facilitating the occupation of the lands along the rail lines.

But the railroad could not solve all the problems of supplying the raw materials indispensible to farming on the prairies. In particular, it could not transport the materials necessary to closing off fields to protect herds from wild animals and Indians. The absence of wood, water, and stone made too costly traditional solutions: the low

stone walls of New England, hedges, or picturesque rail
fences. Earthworks, at one time contemplated, proved
equally inefficient. For these reasons Glidden's invention
represented a veritable revolution in the agrarian
exploitation of the prairie.

In the extreme climatic conditions of the West, barbed
wire is truly efficient. Contrary to naked wire (a single
strand of wire), it resists heat; its torsion eliminates the
slackening caused by dilation; and it is much more difficult to bend or break than naked wire. The firmly fixed
barbs dissuade cattle from forcing their way through the
fences. Other advantages of barbed wire are "light
weight, a great aid in transporting it and in construction;
its universal application and unceasing variety of uses; its
ease of installation; its durability, once in place; and its
adaptability to all sorts of uses."[3]

Above all, barbed wire's low production cost
accounted for its success. When Glidden patented barbed
wire, he also patented a machine to mass-produce it,
which he began doing in 1874 in his factory in De Kalb.
Barbed wire sold for about $4.50 a pound, and twenty
years later for 45¢ a pound. An 1871 report from the
Department of Agriculture estimated that the cost of

The herd protected. "Cattle on the Mack Downey Ranch
on the South Loup River near Georgetown, Custer County,
Nebraska." Solomon D. Butcher, photographer, 1903.
(*Nebraska State Historical Society*)

fencing off 198 acres in the Far West was $640, or $3.25 per acre. In 1897, Glidden's barbed wire cost a settler $2 to fence in an acre, and soon the price dropped even more. As a result, the production of barbed wire shot up from 270 tons in 1875 to 135,000 tons in 1901. (The American Steel and Wire Company took over production, succeeded by United States Steel.) Barbed wire became so important in American history that one historian has written, "it was neither the railroads nor land settlement laws which enabled the farmer to advance beyond the Missouri; it was barbed wire."[4]

But the space west of the Missouri was not empty. Though the gold rush had accelerated settlement west of the Rocky Mountains, the Great Plains between the Rockies and the Missouri River still belonged to the last free and nomadic Indians of the United States.

The Frontier Advances; the Indians Retreat

At the end of the seventeenth century, the population of the thirteen English colonies exceeded two million. At the same time, the Indians experienced a major renewal after years of territorial loss and population decline. Recognizing that the colonies threatened their existence, they organized to protect themselves. One sign of their shift from separate and hostile tribes to united blocs was the creation of confederations: in the Great Lakes region, the Iroquois Confederacy in the mid-seventeenth century; later, and farther south, the Creek Confederation. A clear opposition between the two populations had also become obvious: on one side were the European colonies, which in 1776 became an independent and powerful state; and on the other side, strong Indian nations aware of the dangers and difficulties they had to confront. The area where these two entities met could be considered a frontier, a distinct line of demarcation; but more than a simple territorial division, this area separated two civilizations, two worlds.

At the beginning of the nineteenth century, the Appalachian Mountains formed that "frontier." Forty years later, it was the Mississippi River. The Jeffersonian policy of "Do it, or else" was the basis of this westward expansion. As Thomas Jefferson wrote in a letter, "In the long run, either the Indians assimilate and become citizens of the United States, or they settle West of the Mississippi. The first solution would undoubtedly constitute the happiest chapter of their history."[5]

As the United States grew stronger, the Indian became

a victim, losing his tribal lands under policies intended to be "humane." The government based this treatment on its recognition of the difficulties—possibly fatal—the Indian would experience in integrating himself into American society. Giving protected and reserved lands to the tribes theoretically solved these dangers and difficulties. Thus, legally and in good conscience (after all, those who conceived this solution did so to save the Indians from the white man), the federal government moved whole populations across vast distances and settled them on inhospitable land occupied by other, often hostile tribes.

At the end of the century, westward expansion, accelerated mainly by the railroads and barbed-wire enclosures, ended the Indians' existence as nations and their resistance to the white man. The next task was transforming them into American citizens. As long as a tribe's sense of community stayed strong, the tribe could resist the Indian's new and forced status as a private individual. For this reason, the tribe had to be obliterated. The Indian could be an individual integrated into American society only when his tribe no longer existed. The project of the reformers, humane people concerned about the

Indians' fate, was intended to change them into farmers who fully believed in the concept of private property.

The keystone of this project was the 1887 Dawes Act (a sort of forced Homestead Act), which authorized the president of the United States to parcel out the Indians' lands without consulting them. Each Indian family received two hundred acres of reservation land. The rest was set aside for white farmers. This was a doubly efficient solution, since it simultaneously helped destroy the foundations of Indian society (seen as vestiges of the past and a permanent insult to a modern age based on productivity), and opened to cultivation Indian land unfarmed before the Dawes Act. The white men regarded Indians as squandering the land's potential because they did not exploit it. "Before the white man's arrival, the Indians' land was common property to be shared by all. They could not understand the idea that it could belong to one person. . . . The tribe lived in a territory which met its needs, and on which frontiers and land enclosures were inconceivable."[6]

After implementing measures which threatened to physically eliminate the Indian, the United States turned its Indian policy to cultural destruction: in other words, ethnocide. William Jones, commissioner of Indian affairs between 1897 and 1904, "thought that the system of parceling out land was the 'solution to the Indian problem' and called the Dawes Act 'a powerful pulverizing machine for the destruction of the tribal mass.' "[7] The Dawes Act created a trend. To preserve their way of life, many tribes refused to accept their land parcels, which were then put up for sale. These "hard heads" usually withdrew to the poor and undesirable parts of their reservations. "The Indians really didn't like the idea of enclosed lands. When they saw how barbed wired threatened to encircle their camps and cut off their hunting grounds, they completely rejected life within enclosed land. Their traditional warfare was rendered impossible (surprise attacks and hand to hand fighting). As more and more barbed wire appeared on the prairie, repaired almost as soon as it was cut, the Indians had no choice but to move elsewhere. Rarely were they seen where there were white men and their fences."[8] Indians who accepted the Dawes Act and its system of land parcels usually failed, demoralized by the disappearance of the tribal community, disadvantaged by the poor quality of their land, deprived of all capital, and crushed by the ferocious and often dishonest competition of the white man.

The new western landscape, made possible by barbed wire.
View across agricultural field towards Frank E. Sweet's Ranch
(residence) near Carbondale, Colorado, reached via Colorado
Midland Railway. Circa 1910 (*Denver Public Library*)

"Fence posts and barbed wire around pasture and grazing lands. Near Granby, Colorado." Marion Post Walcott, photographer, September 1941. Farm Security Administration—Office of War Information Photograph Collection. (*Library of Congress*)

The Indians rarely had the desire or the means to fence off their plots, but the white farmers and ranchers relied heavily on barbed wire to protect the land they either owned or rented. Sometimes their lands overflowed onto stretches of unfenced Indian land. In 1903, a northern plains rancher, Ed Lemmon, rented and enclosed 864,800 acres on the Sioux reserve at Standing Rock.[9] Although this is an extreme example, it does illustrate how the Indians gradually lost their land and were surrounded by fenced lands owed by a people who, one way or another, wanted to see them disappear. "From this point on, the Indians were no longer living in merely enclosed space. On the reservations, they were in a fragmented space, under military surveillance which was intended to immobilize and atomize them. All exits from the reservations had be negotiated on an individual basis. . . . On their reservations . . . the Indians found themselves scattered in the midst of a community which encircled them not only with fences but also with hostility and suspicion. This was prison by fragmentation, in which even the fragments were imprisoned."[10]

Barbed wire affected the Indians in two ways. On the one hand, once the barbed wire closed off their lands, the white man could occupy and exploit them. On the other, by the end of the century, barbed wire was used to parcel off the few remaining Indian lands. It chopped space into little bits and broke up the communal structure of Indian society. Barbed wire made the Indians' geographical and social environment hostile to them, so that it became a foreign territory where the tribal way of life was unimaginable and where nomadic wandering and hunting were impossible. In short, it created the conditions for the physical and cultural disappearance of the Indian.

Barbed Wire, the Mythical West, and the Western

The West became an American myth at the very moment when the "frontier," cowboys, the open range, and free Indians had disappeared. There are several possible interpretations of the relationship between the reality of the frontier and the subsequent legend of the West. It probably expresses the sense of loss for a particular society whose originality was rooted in the dynamics of conquest and in its contact with the unknown. This sense of loss quickly became a myth. "The West was more a form of society than a geographical territory. The term is applied to a region where the concept of free land changed

"Settlers taking the law into their own hands—cutting fence on old Brighton Ranch." Solomon D. Butcher, photographer, circa 1900. (*Nebraska State Historical Society*)

"Barbed wire and barbed-wire fence covered by drifting
sand, with tumbleweed growing up all around. This is in a
section which in one day is reputed to reproduce forty
bushels of wheat to the acre. Near Syracuse, Kansas."
Russell Lee, photographer, September, 1939. Farm
Security Administration—Office of War Information
Photograph Collection. (*Library of Congress*)

ancient ways of life and ideas. This transformation created a new milieu which offered opportunity for all, but which also broke traditions, creating new activities and ideals."[11]

Whatever the merits of attempts to link a specifically American sense of egalitarianism with the bloody conquest of the West, it is true that the region did produce certain original but ephemeral traits. Those characteristics lasted for only a few decades, between the end of the Civil War and the dawn of the twentieth century.

When the empire of the cattle barons began in the 1860s, the ranchers set up "laws" to govern the West and to "protect" it from strangers, mainly the farmer-settlers. The principal law on which all others rested was the one that established the open range, "the unwritten law of free access to grass and water . . . [that] prohibited anyone from obstructing the rancher on his way to the prairie or blocking his route to towns and to his fields, and above all to his pastures."[12] Was this an egalitarian vision of society? Perhaps. In fact, the "unwritten law of the open range" established a monopoly. Homesteaders had fenced in the prairie as early as the 1870s, and the cattle empire, founded on free grazing, had largely collapsed.

The battles leading to the defeat of the cattle barons—
the "barbed-wire wars"—are depicted in several Western
films. These movies portray the founders of the West—
strong men, heroic ranchers—in conflict with farmers
and sheep ranchers who monopolize the land and
exhaust it through intensive exploitation. Hollywood
later depicted the ranchers as being in cahoots with
"judges, politicians, and men of the East. They become a
group of inhuman capitalists (e.g., *Heaven's Gate* by
Michael Cimino, in which the ranchers order the murder
of every man in a single county). American society thus
goes from admiring the self-made man to [seeing him as
a villain in] the antitrust struggle."[13] The myth of the
West and its cinematic version also mirror American fan-
tasies, which range from a search for a heroic and
allegedly noble and egalitarian beginning, to regret at the
loss of that heroic origin, or, worse, to the suspicion of its
nonexistence (perhaps the "hero founders" were never
anything more than greedy exploiters).

When the open range disappeared, so did the lone
horseman riding over the plains: the cowboy. Although in
reality he was nothing more than a cow herder, the myth
gave the cowboy all the attributes of the superhero. "For

"Man resting on sack of grain near barbed wire, New Roads, Louisiana." Russell Lee, photographer, November 1938. Farm Security Administration—Office of War Information Photograph Collection. (*Library of Congress*)

the Americans, this union of horse and man feeds a deep nostalgia for mobility and the nomadic way of life. This nostalgia is also the reaction to the uprooting to which modern civilization condemns us."[14] Because the cowboy resisted the culture of the eastern United States and the advance of industrialization, he naturally "became . . . a determined enemy of those sedentary farmers who wanted to divide up the prairie by fencing it in. Kirk Douglas fights against the loss of the open land, and embodies the rough-hewn, unpolished cowboy in *Man Without a Star*."[15]

Barbed wire plays a central role in the film. Dempsey Rae (Kirk Douglas), the archetypical cowboy hero, leaves Texas because he cannot bear the closing off of the prairie and because his brother has been killed in a quarrel over barbed wire. Hideous scars, caused by the "devil's rope," cover his body. He finds work in Wyoming, where the prairie is still open. One night, a farmer invites Rae to dinner. During the meal, the farmer boasts that he has solved the wintertime absence of grass: his remedy is to set aside pastures during the spring and summer. When someone else at the meal asks the farmer how he plans to keep the cattle and sheep away from those pastures, he holds up a piece of barbed wire. The proud cowboy stiffens; his eyes darken: he is infuriated by the sight of the wire. "What's the matter?" asks the farmer. Rae answers contemptuously: "I don't like it, or what it's used for," and storms out. Shortly after, his young sidekick asks why he left Texas. "Isn't there open prairie in Texas which you can see to the horizon?" "Sure," replies Rae, "But they've put in the barbed wire. Before, it was all open, as far as the eye could see, as far as a man could ride or drive his herd. There was nothing to stop him, no obstacles."[16]

According to the myth, particularly its cinematic version, the cowboy is the prototype of the American man, or what he should be. Every American can identify with this heroic figure, dreaming that he is still that same courageous founder of a radically new world, and that he is a free man. The end of the mythic and epic backdrop of the Far West is linked to the loss of three fundamentally Indian values: open space, nomadism, and egalitarianism. Is it not astounding that the heroic figures of the Western myth take their defining qualities from those whom they fought and destroyed? If barbed wire put the cowboy out of work, it also helped destroy the Indian's social structure. The collective memory of the United States should celebrate the Indian as the hero

***The landscape is severed in two by a barbed-
wire fence.*** View of Long's Peak near Estes Park,
Colorado. (*Denver Public Library*)

of an unshakeable resistance to modern productivity, and not the cowboy, the employee of the cattle ranchers. Perhaps the stubborn Indian lies at the heart of the Americans' sense of the lost prairie. Possibly they regret a missed opportunity.

THE FIRST WORLD WAR:
THE TRENCHES

Modern War, Trenches, and
Secondary Defenses

The trench-shelter, used during the first worldwide conflict, is a fortified defensive structure, conceived and built
by military engineers. Unlike permanent fortifications
(city walls, citadels, small forts), trenches are by nature
temporary, or "semitemporary." They are used when an
enemy's attack loses its momentum, slowing down or
stopping. More often they reinforce a defensive position.
In the seventeenth century, temporary fortifications had
little tactical importance, since victory was achieved
mainly by movement and speed. "The pick and shovel . . .

are the recourse of the weak or of those willing to risk nothing."[17] These despised tools became essential when modern rapid-fire rifles and long-range artillery appeared on the battlefield.

During the Civil War the face of modern warfare took shape, seen principally in the widespread tactic of lines of entrenched riflemen firing on the attacking enemy for as long as possible, and in the use of artillery behind the lines of riflemen. This was made possible by arranging a series of defensive items—slopes, sharpened stakes, wolf holes,[18] and above all, chopped-down trees[19]—in front of the trenches facing the enemy, all intended to slow down the attacking force so that he was forced to expose himself. Thus, "a simple trench, defended by two ranks of infantry, covered by trees or other obstacles, and placed on terrain which enables the rifleman to exploit the range of his new weapons, is impregnable and can be taken only by surprise."[20]

The classic trench was a ditch deep enough to protect a line of soldiers up to their shoulders. A parapet facing the line of fire was erected, made of either earth or any other protective material; the defenders could shoot through loopholes in the parapet. Beyond the trench, a gentle downward slope enabled the rifleman to see over the battlefield, so that he was not hindered by his own defenses and so that he could easily kill those who were entrapped in the defensive devices. Of these devices, one military theorist wrote, "one can say that secondary defenses, far from diminishing in importance, in fact increase greatly: and since networks of wire are the most efficient and easiest accessory defenses to install, it will be useful if the battalions load rolls of wire onto their trucks."[21]

The use of wire networks to protect positions was mentioned as early as the Franco-Prussian War of 1870. It soon acquired a respected place in books on military strategy and in bulletins put out by the French Ministry of War. The wire recommended by military experts was not yet of the barbed variety, but rather a single strand of wire. During the Russo-Japanese War of 1904–5, the two sides relied heavily on temporary fortifications, and military theorists saw important lessons in the tactical conduct of that war. Some of them recognized that any future European war would be a confrontation of matériel from entrenched positions. "The Russians had foreseen the need for secondary defenses of the lines at Chiouchampou since war's outbreak. The Russian army engineers carefully put them into place. Multiple and powerful defensive accessories completed these works. . . . [Among them

World War I. Abandoned Russian trench, from a photo
album, circa 1916. (*Archiv für Kunst und Geschichte,
Berlin, courtesy of AKG*)

French trench with barbed wire, World War I. (*Archiv für Kunst und Geschichte, Berlin, courtesy of AKG*)

were] wire entanglements and artificial thorns, 'wolf holes' with and without spikes, all arranged in four staggered rows; the Russians also used electrically detonated mines. In short, all the appropriate defensive items were used, and they gave the Russian positions a formidable appearance."[22]

Immediately after the Russo-Japanese War, wire networks appeared to be the best secondary defense items because they were the most efficient in slowing down the enemy, yet at the same time were barely visible from afar. They were also resistant to artillery fire, economical, and fast to install. A Russian army officer wrote, "The most efficient accessory defenses are wire entanglements; they are more or less impassable and more or less invulnerable to artillery fire. They are the most serious obstacle for an attacking force."[23]

Barbed Wire and the First World War

In 1914, in spite of the lessons of the different "premodern" wars, the French command gambled on its traditional offensive battle plan, that is, movement and piercing enemy lines. As early as September 1914, when the front came to a dead stop, most French trenches were shallow, lacked communication trenches, and had almost no secondary defenses. The first such French defenses were improvised with "wire found in the villages."[24]

The Germans, on the other hand, even though they too based their strategy on aggressive offense, understood the importance of secondary defensive measures better than the French. Their trenches were more solid and even "comfortable." German "trenches equipped with wire obstacles appeared right at the beginning of the war, in Lorraine and in the Ardennes."[25] This led Marshal Foch to comment, "We must imitate the German defenses: deep, efficient, and well protected with wire entanglements."[26]

In spite of the differences between the belligerents' trenches, their secondary defensive systems shared certain similarities. Stakes between five and six and a half feet high were planted at regular intervals of six and a half to ten feet, joined by wire. The depth of the networks had to be one hundred feet to give the defender time to prepare himself. Each line of stakes was set off from the one in front, giving the network, when viewed from above, the appearance of contiguous diamonds. Wire joined the stakes to one another from top to bottom in such a way that space was completely filled. Sometimes

Guard dog. Champagne (France), 1916. Phillipe Larousse-Moreau, photographer. (*Archives Larousse-Giraudon*)

the density of the entanglement was improved by planting small stakes among the big ones.[27] Several kinds of wire were used, most often a mixture of barbed wire (called at that time "artificial bramble") and bare wires of variable thickness. Entanglements consisting entirely of barbed wire were rare, for economic reasons, but barbed wire provided the network with the greatest strength.

The "artificial bramble" used by the army was generally the same as that used to fence in fields. The military variety was distinguished from the agricultural mainly by the density of barbs per meter: seven splinters for the agricultural wire, fourteen for the "strong bramble," nineteen for the "Iroquois" type. And there were wires whose shapes the soldiers saw as particularly useful, such as square wire.[28]

The lightness of barbed wire gave it its tactical importance. It was also hard to see. By day, it made reconnaissance by airplanes and balloons difficult. In addition, it caused infantrymen to make fatal mistakes. Early in the war, "The first men killed never saw the enemy. . . . Where were the bullets coming from? Impossible to know because the enemy trenches were invisible."[29] At night, the soldiers sometimes found enemy trenches only by blundering into their barbed-wire entanglements.[30] Moreover, the networks were camouflaged with vegetation.

Barbed wire's lightness made it immune to enemy artillery fire. Because it is also supple, it would bend rather than break under heavy bombardment. Even broken up, it was still a formidable obstacle. While artillery fire quickly destroyed costly ramparts, the entanglements merely absorbed explosions. A veritable avalanche of artillery shells was necessary to bury the wire over an area sufficiently wide before attacks could be launched.

Barbed wire was economical in all senses of the word. It could be used to make a defensive system one hundred feet deep by hundreds of feet long. The barbed-wire network eliminated the superfluous, doing away with high, thick defensive walls. In their place it left only a fine metallic skeleton.

Barbed wire was easily repaired or replaced. The networks could be worked on at night or in the fog. When the trenches were too close to each other (sometimes less than a hundred meters apart), variations of the classic entanglement were thrown over the parapet. For instance, the soldiers could launch the "Brun" bramble, a sausage-shaped network formed by a series of big meshes of naked or barbed wire. The "sausage" was about four and a half feet in diameter and could be rapidly unrolled; it could even expand itself to about one hundred feet. "Once on the ground, the whole thing made up an unformed, jagged and mangled mass which was almost impassable."[31] The soldiers could also throw out "Friesian horses,"[32] as well as the pliable "Bigot" networks, which, once on the ground, made barbed-wire mattresses on the battlefield.

Although the networks were light, they were highly efficient obstacles. "At 14 hours a bayonet assault on the German trench is scheduled. . . . The soldiers ask themselves what good is the attack? We know we'll never get there, because the barbed wire will hold us up and then we'll all be killed on the spot."[33]

Up to the end of the war, many ways of destroying the wire were devised. The simplest method, of course, was cutting the wire with shears; but this was extremely dangerous, basically suicidal. As the war went on, the soldiers invented ingenious devices to destroy the enemy's wire from a distance. They thought up gadgets like the "wood louse," a land torpedo on wheels, or the "Schneider crocodile," which crawled under the wire and blew it up. They used armored farm tractors and steamrollers, as well as blowpipes filled with oxygen and acetylene and protected by wheelbarrow-shields. The English specialized in developing techniques for getting through barbed-wire brambles, one of which was the Barbed Wire Traversor, a sort of thick, resistant blanket. After it was thrown over the network, soldiers could walk or crawl toward the enemy position. Perhaps the most astonishing invention was the assault torpedo invented by a Lieutenant Mattei. It consisted of a grappling hook attached to a wire thirty feet long; all along the wire, "firecrackers" were attached. The grapple and wire were thrown to land between the enemy's trenches and his wire, and when the snakelike device settled down on the wires, the "firecrackers" were detonated. The explosion opened a corridor in the wire through which the attacking force could assault the enemy trench!

In spite of all these devices, intense shelling by 75-millimeter cannon was the best way to destroy barbed-wire networks—that is, up until the tank appeared. Once the tank arrived, barbed wire became a less useful defensive device, and thick ramparts resumed their defensive role, as well as bunkers built of reinforced concrete.

The Poilus* and Barbed Wire

In the accounts of the war, the novels of soldier-writers, and the letters of humble *poilus,* barbed wire was described as both dangerous and terrifying. First of all, it represented the risk which the soldier must run, in defense as well as attack. Reinforcing the network was particularly deadly, since it sent the soldier out into the open, where he was exposed to stray bullets, bursting shells, or the sharpshooters who specialized in head shots. "When the wires were cut, I had my repair section fix them. . . . In general, the others rushed to get out of a bombarded sector, but the repair section routinely went immediately to the broken wire. Day and night, they visited shell holes, still warm from explosions, to reattach two ends of barbed wire: a task perilous but without military reward."[34]

Above all, barbed wire appalled those who attacked. After surviving the tumult of no-man's-land, they had to reach the enemy's trenches, whether the wire was cut or not, hiding for better or for worse in shell holes. This was

Poilu, "hairy one," is a French slang term for an infantryman of World War I; *poilus* is the plural form. Translator's note.

No-man's-land is not empty. Corpse of a German soldier entangled in barbed wire, World War I. (*Archiv für Kunst und Geschichte, Berlin, courtesy of AKG*)

the procedure for the wire-cutting parties: "Each group was made up as follows: a lieutenant, and behind him, six sappers, without rifles, but armed with shields and enormous shears to cut the wire. . . . Some of them actually reached the barbed wire: alas, it was too thick."[35] Soldiers who were blocked by the bramble were massacred by enemy fire. "The attacks were preceded by artillery barrages which sometimes lasted for days. . . . In spite of the attempt to destroy the wire with barrages, the infantry was often cut down by machine guns, and the soldiers . . . were stranded in the undamaged defensive networks of the enemy."[36]

Barbed wire also became a perennial topic in literary works about the war, forming part of the "aesthetic" of the battlefield. Images of disaster—the torn-up landscape and the mangled bodies—were engraved in the memories of the combatants.[37] The desolate look of the field plowed by artillery shells inspired descriptions which, more than merely deploring the devastation, tended to evoke a sublime, even monstrous modern technology run amok. The craters and mud of no-man's-land, its uprooted trees and its villages destroyed a hundred times over, revealed the essential inhumanity of the industrial world. Modern technology's destructive power overwhelms and stupefies the lone individual. No-man's-land becomes a "work of art" for whoever contemplates and describes it, and barbed wire is an essential element in this nightmare picture. "I look from the parapet. Through the vaporous, dismal, and lurid atmosphere which the meteor has spread, I make out the stakes and even the slender lines of barbed wire crossing each other from one stake to another. To me it all looks like strokes of a pen which scrawl and erase the ghastly and cratered field."[38]

No-man's-land was not empty. It was peopled by the dying, the dead, and pieces of corpses. "Oh yeah, they know some good jokes, all the wise guys who write about the war . . . Dying in the sun . . . By the way, I'd like to see one of them die like an animal, with his mouth open in the barbed wire, and then I'd ask him to admire the land-scape."[39] The corpse stuck in the wire stayed in full view of all the soldiers in the trenches, both friends and ene-mies, so they could see just what was waiting for them. Sometimes this sight so demoralized the soldiers that they risked their lives to unhook their dead comrade. "One of our officers is hanging on the German wires and in the several attempts to bring him back many brave men lost their lives."[40] The image of the dismembered

Prisoners of war. World War I. Phillipe Larousse-Moreau, photographer. (*Archives Larousse-Giraudon*)

body rotting on the wires, as if trapped in a spider-web, showed the *poilu* the absurdity and pathos of his situation. "Hundreds of dead soldiers, many from the 37th Brigade, were scattered like debris from a shipwreck. Most of them had died in the enemy's barbed wire, like fish in a net. They hung there in grotesque postures. Some looked as if they were praying; they had died on their knees and the wire stopped their fall to the ground."[41]

Barbed wire was a salient attribute of the memory of the Great War. It never became a metaphor for the war, because it does not symbolize the entire conflict, or even the fighting in the trenches. Nevertheless, barbed wire could be said to have the "artistic" role of evoking the monstrous sublimity of the forces of destruction liberated by modern war. In the accounts and images of the Great War, barbed wire is significant only as a part of an overall aesthetic. Only after the war did it become a universal symbol of the whole. Its decisive role in the Nazi concentration and extermination camps made barbed wire the symbol of the worst catastrophe of the century.

THE CAMP

Setting Up the Camp

The Nazi camps were not identical to one another. They were not built at the same time, or in the same places, or for the same purposes. Some camps dated from the beginning of the Reich (e.g., Dachau, built in 1933); others were constructed in the middle of the war (e.g., Treblinka, in 1942). Some were isolated, while others were near a town or even a city. The prisoners and the living conditions imposed on them varied considerably, in spite of the absolute and undisputed tyranny of the SS.

The principle of construction of the camps was relatively homogeneous. The KZ of Buchenwald is the classic model.[42] It was filled with rows of barracks, the front of

Auschwitz, electric fence between the blocks. Poland,
1983. Jan Hausbrandt, photographer. (*Courtesy AKG Paris*)

which faced the place for roll call, which in turn faced the entrance gate. The camp was surrounded by a double fence of electrified barbed wire thirteen feet high. The enclosure was under constant surveillance by watchtowers situated every seventy-eight feet outside the enclosure. The guards were equipped with automatic weapons and powerful spotlights pointed at the enclosure. The camp administration (the Kommandantur) and housing for the SS troops were outside but near the camp. This same layout—the prisoners' barracks, the double fences of electrified barbed wire and guard towers—was found in all the Nazi camps. It constituted the typical concentration-camp landscape.

The central element of the camp's architecture was the barbed-wire fence. The fence was usually the first structure built when the camp was set up. Certainly, sometimes camp buildings were constructed at the same time as the fences. Thus, at Belzec, "whereas we, the Poles, built the barracks, the blacks [Ukrainians] put up the stakes and barbed wire."[43] But the camp was not a camp properly so called until it was surrounded by barbed wire. "The installation of the electrically charged barbed-wire fences ended the first stage of the construction of Buchenwald's exterior."[44]

The primacy of the fence was based on the model of the camp at Maidanek. In the autumn of 1941, the Nazis began to build this vast concentration camp (where later prisoners would be gassed). For all of the first phase, "five thousand Soviet soldiers . . . were penned up like cattle in an enormous field surrounded by barbed wire," with almost no shelter.[45] Each day, hundreds of men died, so that by the end of November 1941 only fifteen hundred were left alive. Such was the concentration camp without any of its usual buildings: a place organized for absolute destitution, where men were transformed into forsaken cattle. "The human masses shut up [in the camps] were treated as if they no longer existed, as if what had happened to them was of no interest to anyone, as if their deaths had already been determined. It was as if an evil spirit, seized by madness, was amusing itself by keeping the prisoners between life and death before letting them go to their eternal peace."[46] This was the essence of the camp: barbed wire.[47]

None of the camps were built to last. Even an immense one was constructed in such a way that it could disappear from sight and memory. It was there, but it was not there. It was transient. It resembled a tent, which can be put up and taken down the next day. "Buchenwald is a chaotic

The barbed-wire perimeter and watchtower of the
Buchenwald concentration camp, built in 1937 in
Ettensburg, near Weimar, Germany; prison camp of the
Russian occupying army, 1945–1950; memorial since
1958. Dieter E. Hoppe, photographer, 1994. (*Courtesy
AKG London*)

city, a sort of unfinished capital city, something like an encampment, with its quarters hastily and summarily planted and its swarming humanity."[48] The importance and utility of barbed wire are obvious. It is the best material for a temporary structure. A solid wall leaves traces, but a barbed-wire fence leaves nothing. This is why it is so difficult to find evidence of several camps which were completely razed at the end of the war. At Sobibor, "as at Belzec, the terrain was plowed and planted with trees in order to hide the traces of extermination."[49]

Barbed wire was thus used in the camps for reasons of economy: it is cheap, easily installed and dismantled, and also deadly when electrified. However, this "burning frontier"[50] had specific effects on those inside it. Barbed wire was not just a material used in the camp. It provided the essential foundation of the totalitarian management of space.

Inside the Wire: Organized Desolation

The monstrosity of the camps can be introduced in two distinct but complementary ways. One way is to see the camps as dungeons, places of the most unimaginable tortures. Another is to consider them as cities of the most radically totalitarian society. If the camps are viewed as places of extreme cruelty, their architecture does not matter. What is important is the unimaginably horrible acts that occurred there, not the barbed wire and everything else that physically constituted the camps. If the camps are perceived as cities of a totalitarian society, the political use of them is significant. They were not black holes, but rather the physical realization of the totalitarian dream: a society of total domination. In this case, the camps' architecture matters greatly. It constitutes the totalitarian organization of the whole complex.[51]

In the camps, barbed wire organized space and defined the ranks of the camp hierarchy. The fences radically separated the camps from "normal society." "Everywhere [was] the sinister tight iron grip. We never saw where the barbed-wire fences ended, but we felt their malign presence which separated us from the world."[52] Anyone entering the camp had to forget what he knew before. If a prisoner wanted to survive, he had to quickly understand that his former behavior outside the camp had no value. Here, another sort of money was in circulation. He had to understand that here, in the camp, "Everything is possible."

Auschwitz. Prisoners after liberation, 1945; the watchtower
looms above them. (*Courtesy AKG Paris*)

Primo Levi recounts how on the day of his arrival at Auschwitz he tried to quench his burning thirst by sucking a piece of ice detached from the edge of a window. A kapo rushed over to him and ripped it from his hands. When Levi asked the kapo why ("Warum"), the kapo's answer summed up that other world that was the camp: "Hier ist kein warum," or "Here there is no why."[53] Similarly, at Buchenwald, a group of newly arrived prisoners asked their kapo about the health of a comrade who had been chosen for a certainly fatal transport convoy. Without intending any malice, the kapo laughingly answered, "Here there aren't any sick people. There are only the living and the dead."[54]

Teaching a new prisoner that "everything is possible" began always with "Here," that is, inside the barbed wire. The prisoners had to explain to someone who had just arrived in the camp, and who still had the flesh and face of a free human being, how the all-powerful SS would soon rob him of his identity. Once he was behind the barbed wire, he was no longer a human; he was lower than an animal. He was a mere body, a head *(Kopf)*, a piece *(Stück)*, who at best would die slowly.

Beyond the barbed wire, between the camp and the world, was the SS. Its members stayed out of the camp whenever they could. Camp organization, surveillance, and discipline were undertaken by those prisoners who formed the camp aristocracy with its many ranks: head of camp, block chief, room chief, kapo. "The SS apparatus was completely outside the camp. The SS controlled the routes which led to the concentration-camp universe. In the gaps of the pine trees rose the guard tower and its machine guns. Beside the trees, on the edge of the road, were the barbed-wire fences. . . . The SS guarded the entrances and counted the men."[55]

Inside the camp, barbed wire marked off places which had a special status. Space was organized in such a way as to render arbitrary classifications visible. Women were separated from men. Certain nationalities were isolated, in particular the Soviet prisoners of war. Jews selected for work sometimes had a section, as did auxiliary SS troops—Ukrainians, for example.

The internal separations also created sectors where the living conditions were "improved." The *Revier** was almost systematically separated by a barbed-wire fence, often electrified. In most of the camps, a stay in the infirmary amounted to respite and sometimes to survival, if

---

*A German word for an ersatz hospital. Translator's note.

only temporary. On the other hand, in the extermination camps, at Auschwitz in particular, the infirmary was a particularly dangerous place, since there routine selections sent the dying to the gas chamber. Experiments conducted by Nazi doctors were also carried out in the infirmary.

Barbed wire marked the passage to a new circle of hell. At Buchenwald, a prisoner entered that circle when he went into the "little camp." This "little camp" was made up of tents and surrounded by wire fences. Here selected prisoners were confined in dreadful conditions where there was almost no sanitary equipment. Even inside the "little camp" there was a hierarchy of the horrible. A prisoner could be thrown into the "rose garden," a cage made up entirely of barbed wire. This was barbed wire inside barbed wire inside barbed wire—Chinese boxes of barbed wire. "It was there [inside the cages] that certain victims died of hunger, where it was 2 degrees at night and 20 degrees in the day. They died under the gaze of their comrades who did not know when their turn would come."[56]

The gas chambers and crematoriums had their own barbed-wire fences. To shroud everything in secrecy in the extermination camps, the barbed wire was braided

The passage to a new circle of hell. Concentration camp at Elmsland, Germany—guard in watchtower outside barbed-wire fence, circa 1935. Walter Talbott, photographer. (*Library of Congress, Prints and Photographs collection*)

The political management of space. Soldiers erect barbed-wire barriers, Austria, February 1934. (*Courtesy AKG London*)

with branches so that the gas chambers were invisible. At Sobibor and Treblinka, the victims went through a narrow barbed-wire passage also braided with branches, to arrive at the gas chambers disguised as showers. "We received the order to go with the Germans. They escorted us to the gas chamber, situated in the second part of the camp. 'Is it far?' 'No, not very far.' But everything was camouflaged. The barbed wire was covered with branches so that no one could see, never imagining that the passage led to the gas chambers. 'Is this what the SS calls *the passage?*' 'No,' they would say. 'It's the road to Heaven.' "[57]

Barbed Wire: The Symbol of Extreme Captivity

Imagine a close-up of a piece of barbed wire. Do you think of a fence for a field? No, of course not. By itself, barbed wire is nothing more than an agricultural tool. To imagine a field, we think of a cow or a sheep. On the other hand, to visualize confinement, seeing a prisoner is unnecessary. A picture of barbed wire alone is enough to evoke the concentration camp's prisoner and oppression. It has become a graphic symbol for incarceration and political violence. It links three modern disasters: Indian

ethnocide, the butchery of the First World War, and Nazi
extermination. After Auschwitz was liberated, Primo
Levi put it this way: "Liberty. The breach in the barbed
wire gave us a concrete image of it."[58]

During the immediate postwar years, the French associated barbed wire with the trauma of the occupation. In
1946, people going to the Bois de Boulogne in Paris
angrily objected to the French counterespionage organization's moving into buildings on Boulevard Suchet*
which had been occupied by the German navy. The
Kriegsmarine had surrounded these buildings with
barbed wire. "These barbed-wire fences, this wall, bring
back painful memories. We love to amble around Paris.
We accept dead-end streets as a law of urban nature.
What we cannot get used to is the barbed wire around the
Bois de Boulogne. Its presence shocks us. The buildings
of [the French counterespionage service] might need protection, but not in the manner of their former occupants.
And we Parisians, who love the Bois, protest that the one-
way streets set up by the Germans are still in place."[59]

*The Boulevard Suchet runs along the Bois de Boulogne.
Translator's note.

The returned deportees and prisoners of war adopted barbed wire as a sign of their special identity, their activities, or their preoccupations. Right after the war, life for those who returned from captivity was difficult. The stigma of defeat injured the dignity of the ex-POWs. The memory of the concentration camp relentlessly haunted the former prisoners, isolating them from the rest of the French population. Often it was difficult for these "returnees" to find work or even basic necessities. For all of these reasons, some of them wore "a miniature piece of barbed wire" in their lapels.[60] The badge distinguished them from other Frenchmen.

These returnees communicated their experiences and their political consequences in print: *Le Front de Barbelé* [*The Barbed-Wire Front*] was a weekly magazine put out by the Association of Prisoners of War of the Seine and first appeared in 1945; *Après les barbelés* [*After Barbed Wire*] was the monthly information publication of the Association of the Lower Rhine Prisoners of War (1948); *Les Barbelés* [*The Barbed Wires*] was the monthly bulletin of the Associations of Prisoners of War, 1939–40; *Cisailles et barbelés* [*Shears and Barbed Wire*], the magazine of the National Union of Escapees of War. This need to communicate was also expressed in plays and nonfictional accounts: *Dans les barbelés* [*Inside the Barbed Wire*], a drama about captivity written by the prisoners of Colroy-la-Grande in 1945, and *Barbelés sanglants* [*Bleeding Barbed Wire*] by Richard Gueutal, published by the Association of Ex-Prisoners of Stalag VA (1948).

Books about the camps often used the image of barbed wire on their covers. Sometimes it was a photo of the barbed-wire fence, far more effective than a picture of the whole camp.[61] Some book jackets used merely a stylized representation of a few bits of barbed wire.[62] For instance, the cover of Annette Wievorka's *Déportation et Génocide. Entre la mémoire et l'oubli* [*Deportation and Genocide: Between Remembering and Forgetting*] is white in the middle with a single piece of stylized barbed wire. Andrej Kaminski's history of concentration camps since 1896 is presented with four pieces of barbed wire on a red background.[63]

Barbed wire thus became an almost universal symbol of the camps and, more generally, of fascist and totalitarian violence, because of its function and its powerful evocative capacity. Its form illustrates its function. It is a line which demarcates space, locks it up, and, like prison bars, immediately invokes loss of liberty. Lines and

Government troops seal off the Spittelmarkt with barbed
wire to protect the city center from a Spartacist attack.
Berlin 1919. (*Archiv für Kunst und Geschichte, Berlin,
courtesy AKG London*)

World War II, Eastern Front. German Army soldiers
at a barbed-wire barrier, winter 1942. Alois Beck,
photographer. (*Courtesy AKG Paris*)

points, bars and knives—they directly express the violent and oppressive purpose of barbed wire.

Barbed wire was such a powerful and persuasive symbol that without it, a camp was not perceived as a camp. This was useful to anyone who did not want to believe that the camps existed. In his memoirs, Joë Nordmann recounts how he did not believe Margarette Buber-Neumann's testimony about the Soviet camps during the Kravtchenko trial. That a camp could be twice the size of Denmark was "incomprehensible" to those who blindly believed in the regime. "But what was she talking about? Was it one of Stalin's concentration camps? If it had no walls or barbed wire, and if people moved around in it as they pleased, how could it have been a camp?"[64]

2

BARBED WIRE

AND

THE POLITICAL

MANAGEMENT

OF SPACE

Barbed wire fulfilled a preexisting need to delimit space; it was nothing more than a tool. The agricultural function of a fence is to defend arable land from predators. The farmer encloses his land to prevent intrusions by wild beasts, herds of domestic animals, or thieves, all of which threaten his field, his possessions, or him. Thus, it creates two polarized spaces: a threatening exterior and a protective interior. Since it protects, it reassures the people occupying the interior. Enclosing is therefore a political act, because it marks out the boundaries of private property, assists in the effective management of land, and makes social distinctions concrete.

Dividing space with enclosures has both a static and a dynamic component. The static one is the material presence of the fence. Here the enclosure announces ownership or indicates the particular status of the enclosed space. It does not create land divisions; it only designates them. In contrast, a fence's dynamic component is its capacity to produce a difference in space; that is, its power to chase off intruders. In this case, an enclosure is both a sign and an action.

Barbed wire came into existence in the American West because of a need to declare property ownership and to exclude intruders. Because the western lands to be occupied were immense, a fence which repelled but was as light as possible was necessary. In satisfying these needs, barbed wire had decisive political effects: its use produced masses of rejected animals and men. Its lightness also reinforced the relationship between spatial delimitation and surveillance, a connection which has important present-day political consequences.[1]

A FRONTIER BETWEEN
LIFE AND DEATH

Barbed wire excludes and includes. Its function is always to magnify differences between the inside and the outside. Barbed wire was added to preexisting elements in order to enhance separation. Thus, it was a supplementary element in protecting fields and herds, in separating whites and Indians, in defending trenches, and in imprisoning.

How, then, did barbed wire, which began as a simple auxiliary tool, become the essential element in the frontier between life and death? How did it acquire its role in the major modern processes of separation?

The Indians, the German and French soldiers, and the concentration-camp prisoners were not simply excluded by society and then pushed toward its edges. Spatial delimitation itself does not create second-class citizens. Those who are rejected and expelled beyond barbed wire, or imprisoned inside it, do not face difficult lives, but violent death.

According to Tocqueville, the dynamic of radical exclusion of which the Indians were victims is the negative face of an equally radical inclusion. Democratic American society is supposedly based on the equality of its citizens. Anyone willing to "play the game" of citizenship, as it is defined by the Constitution, theoretically has equal rights. In the United States, this dynamic of inclusion was a new power, a universally believed principle. It was a force unleashed on vast and untouched territories which had no historical obstacles to bar its momentum.

Contrary to what the advancing white Americans believed, the West was not virgin; it was already inhabited by a civilization with its own values and politics. But no meeting was possible between the Universal conqueror and the welcoming, individual Indian. The Indians' only "solution" was to withdraw as the march of the "Universal" advanced. Because the continent is finite,

Japanese prisoners of war in a camp in Guam,
after receiving news about the surrender of
Emperor Hirohito, August 1, 1945. (*Courtesy
AKG Paris*)

when the Indians could retreat no farther, they died.[2] The process of democratic and liberal inclusion, precisely because it saw itself as universal, covered a huge territory, but at its edge was a bottomless pit. "It seemed as if the inclusive movement, by which democratic civilization constitutes itself, also drew a limit from which the inclusive energy recoiled." This withdrawal pushed the Indians to the abyss.[3]

However, the nature of the division did not rest solely on the legal difference between white men and the Indians. The Indians were destined to die not just because they were unprotected by democratic law; they existed outside a zone where the lives of men (and of animals) were managed, enhanced, and stimulated by a complex of technologies of power which Michel Foucault called "biopolitics." At the end of the eighteenth century, governing powers began to concern themselves with demography, health, and environment, partly in response to the growing importance of the quantity and quality of industrial labor. This was a decisive change: as monarchical power was dying, another power developed, one "destined to produce forces, to grow and regulate them, rather than to enchain them, bend them or destroy them."[4]

Barbed wire was overwhelmingly successful in the United States partly because it coincided with the country's biopolitical needs. At first, the farmers needed fences to mark off their property and to defend it against the exterior, especially against free-ranging cattle. They needed a system which neither injured the animals nor angered their owners. The goal of the early manufacturers of barbed wire was preventing herds from entering closed land. The first wires were almost invisible to the cattle, and their long, sharp barbs wounded many animals. In the great prairie, where cattle wandered untended, these injuries were often fatal. "Viewing the problem from a new angle, the manufacturers produced barbed wire which the cattle could easily see rather than wire which inflicted pain. Although this conflicted with the principles of barbed wire, the new product was put on the market. Several models of less injurious and more visible wire were now developed."[5]

Barbed wire was thus adaptable to several uses, however contradictory: although the manufacturers varied its form and thickness, it kept its capacity to drive away but reduced the damage to the repelled animal. This is why the cattle ranchers also turned to wire to provide a protective interior for their herds, a much more fertile area than the dangerous prairie outside the wire. Barbed wire's capacity to repel without wounding sparked off the rapid development of cattle raising and land cultivation, two activities which must fall within a sphere of "bio-political comfort."

But barbed wire's repellent efficiency has two faces: one cannot protect one side without threatening the other, and the repelled undesirable is forced toward a harsh and even desolate exterior. Giorgio Agamben writes, "[I]n any modern state, there exists a point which marks the moment when the decision on life is transformed into a decision of death, and when biopolitics . . . becomes thanatopolitics*."[6] Barbed wire clearly was, right from its inception, an active and efficient tool in determining who should live and who should die.

The American frontier's reality was completely different from its myth. It was not the limit between light and darkness, civilization and barbarianism; instead, it was the division between whites who were protected because they worked the land and Indians exposed because they

---

*"Thanatopolitics," from the Greek *thanatos* for death, and "politics," thus the politics of death; here it is opposed to "biopolitics," from the Greek *bios* for life and "politics." Translator's note.

did not. In this sense, barbed wire was a pioneering element in capitalism and industrial progress because it helped destroy the society of people who rejected both.

Two other historical events already discussed in this study produced the evolution from biopolitics to thanatopolitics: World War I and the Nazi concentration camps. The First World War brought about two changes. On the one hand, the foreigner, whether German or French, stopped being "someone other than oneself" and became a monster or the absolute enemy. The nature of modern war radicalized the conflict and transformed it into a struggle for life, in which the enemy had to be not just destroyed but literally pulverized. The barbed-wire front protected the country not from defeat, but from destruction. On the other hand, as essential as the front-line soldiers were to the country's survival, they nevertheless were isolated from the rest of society.

In the camps, governed by thanatopolitics, men's deaths were indistinguishable from their humiliation and their dehumanization.

Among the modern political devices of separation, barbed wire becomes a tool of extreme polarization. On one side, the interior, rights are protected, and life is bettered. On the other, the exterior, the arbitrary is

unleashed, people accelerate destruction, and they manu-
facture death.

Along the thread of these three historical formations,
it is possible to see an intensification of the brutality of
power, of killing in the "exterior," defined by barbed
wire. This could, of course, be attributed to the perfection
of the tools of destruction and to the differences among
the political situations of the Indian, the soldier, and the
concentration camp victim. But this would disregard
another important factor—that of geometry. That is, in
these three cases, the "exterior" produced by barbed wire
did not have the same form.

In the United States, the frontier was a line advancing
toward the West, which the Indians resisted but from
which they finally retreated. The violence of the frontier
was reduced by the possibility of escape to the backcoun-
try. When this possibility disappeared and flight was no
longer possible, the bell tolled for the Indian.

During the First World War, two armies faced each
other, equipped with all of the means of destruction
which the modern state can provide. The violence was
concentrated in no-man's-land, a relatively narrow strip
which belonged to no one. However, lines of flight
existed. Soldiers who survived the horrors of no-man's-

land could always return to their own trenches by crossing over their own barbed wire. They could also try to break the stalemate of trench warfare by using the primordial tactic of the flank attack. Ernst Jünger recounts how, with a few comrades, he organized commando attacks on the English trenches. Jünger's small elite groups would slip into the enemy trenches and run through them, tossing grenades right and left.[7]

In the case of the camp, the geometrical form which set the scene for violence was attained by sealing the "exterior." To speak of the "exterior" when referring to the space inside the concentration-camp barbed wire is not contradictory: it was the place/nonplace of death's absolute and arbitrary tyranny. No escape was possible. From all directions barbed wire imposed the forms of brutality devised by the SS: hunger, thirst, cold, beatings. In this interior/exterior of barbed wire, death was concentrated. It was in the very air one breathed, saturated with the odor of burning bodies.

GUARD THE HERD
AND KILL THE BEAST

Barbed wire is intended for living beings. Wooden fences bar space, but they are not specifically conceived to assault the living being. Boards and posts have a role with respect to space: they separate. On the other hand, barbed wire has an active relationship with bodies: it removes and alienates. But this relationship occurs at the subtlest of levels, that of people's awareness of suffering and their inclination to avoid it. It is not concerned with cattle, sheep, stray dogs, jackals, Indians, cattle thieves, horses and their riders as individual beings; it lumps all together, and in the same way, according to whether they are inside or outside. The interior must be protected from stalking shadows, from migratory masses, and from intruders.

As has been seen in this study, barbed wire is a device which separates those who will live from those who will die. More precisely, it produces a distinction between those who are allowed to retain their humanity and those reduced to mere bodies. On the one side, the productive subjects are preserved and covered in the guise of democratic rights. They are a herd, but one with a human face. On the other side, the abandoned are deprived of rights—they resemble beasts more than humans. They are not the equivalent of a herd, though, which has economic value and belongs to the interior. Those banished to the exterior are lost in the unknown; for those of the interior, they are wild animals, negative and threatening. A vast vocabulary exists to label these beings beyond the barbed wire.

For the white American, the Indians are savages; they are uncivilized men when they are peaceful, and beasts of prey when they go to war to defend themselves. In the West, at the end of the nineteenth century, it was quite common to say that barbed wire was used as a "protection against wolves or wolfish Indians."[8] In classic Western films, the figure of the bestial Indian dominates: "Feathered and painted with glaring paint, no longer looking like humans, but rather half animal, half thing, the Indians yelp, rise up, sometimes they kill (although never the hero), and are killed. . . . They are just obstacles to be removed."[9]

During the First World War, the adversary was not someone to conquer because of geopolitical disagreements. The enemy had to be eliminated because he endangered *la patrie,* the fatherland, the race. France had to be defended against the monstrous Huns, cattle, wolves, dangerous germs. "German officers, no, no, no. My old friend, the Germans are not men, they're monsters. You could call them the vermin of the war. You have to see them up close, hideous, huge, stiff, thin as rails, but also slobs."[10] In the concentration camp, people are scraps, pieces *(Stuck),* cow heads *(Kopf),* pigs, even lice, cockroaches, vermin capable of infecting the clean, healthy Aryan.

Portraying people as beasts made it easier to treat them that way. Prisoners were penned up like animals; Jews were treated like lice. The Nazis had long portrayed the enemy as a noxious animal, an agent of the destruction of the German race. "When a people sees that its entire exis-

New frontiers. Barbed-wire fence in front of the Reichstag, Berlin, 1964. (*Courtesy AKG Paris*)

The barbed-wire enclosed camp for migratory workers
at the Cannon Company of Bridgeville, Delaware.
Jack Delano, photographer, July 1940. Farm Security
Administration—Office of War Information
Photograph Collection. (*Library of Congress*)

tence is threatened by an enemy, it has only one goal: the annihilation of that enemy."[11] Thus, the camps were sustained by animalizing the enemy. "You see [the prisoners], they are animals, and the worst species, as well. You've been told this. They are ugly, they stink, they are weak, they are cowards and they fight each other to eat. No Aryan would do such a thing." Weren't they brought, sometimes naked, to the camps in cattle cars whose skylights were shut with barbed wire?

The simple act of placing men behind barbed wire produces superimposed images of men and beasts. Barbed wire in the camps functioned as a visual apparatus of Nazi propaganda.

The technical polyvalence of barbed wire—its capacity to repel any living thing, whether a cow or a dog—produces a kind of shock when it is used to enclose people, shaking their certitude that they are human. It confirms their fate: like beasts, they are to be worked or slaughtered.

Barbed wire, justified by an animalizing propaganda, is a means by which man is transformed into pure living material, able to be liquidated or forced into labor: at once livestock, savage beast, and vermin. In the camps, it was about guarding the herd, and killing the beast.

BARBED WIRE AND
SURVEILLANCE

Neither a fence nor a wall is ever sufficient in itself. For that, one would need a totally indestructible rampart, which is impossible. All material barriers eventually need repair, improvement, and especially surveillance. Although barbed wire is an effective obstruction, it is also paradoxically vulnerable: its lightness and suppleness make it somewhat fragile. If no one is watching, cutting the wire with a pair of shears is easy. Thus, field fences are guarded by their owners. Lookouts scrutinize no-man's-land from the trenches, aiming their rifles at anyone trying to cut the wire. In the camps, even though the fence was electrified, without the guard towers the camp was not secure. The rare escape attempts show that the main security problem was in the towers, not the fences. At Mathausen, in 1945, some Soviet prisoners of war facing extermination attempted to escape. "The escape began with the assault on one tower. . . . Then they neutralized the electrified wire by causing short circuits with the aid of wet blankets. They captured a second tower with the weapons taken from the first."[12]

The guards in the towers watched neither the interior nor the exterior. Their concentration was focused myopically on the fences. This is why it is not possible to speak of a "panoptical" surveillance with respect to the camps, even if some of them were built to achieve the architectural perfection of Jeremy Bentham's model prison.[13] Bentham, the English utiliarian thinker, conceived the *panopticon* at the end of the eighteenth century to cut costs. A single guard, placed in the middle of a circular building, could observe all the prisoners shut up in their cells in the periphery. Because the prisoners could not see the guard, his presence became almost superfluous. This was virtual surveillance; it was enough for a prisoner to think he might be being watched for him to behave as if he were being watched. Bentham's prison achieved one of modern power's main objectives: to reeducate and discipline the prisoners rather than simply incarcerate them.[14]*

The combination of barbed wire and surveillance is of a different order. Although quite compatible with panoptic control, it is not concerned with the general behavior of those being observed, but simply their position with respect to the boundary which contains them. The combination's sole purpose is preventing people from leaving an authorized area and entering a forbidden one.

Barbed wire and surveillance form a unique tool in the spatial application of power. They constitute one unit not merely because they are connected but because they overlap and are inseparable. The guard holding vigil over barbed wire is also protecting himself. Guards are placed outside the fence: one cannot imagine guard towers inside the concentration camp. Deciding which came first, the tower or the wire, is impossible, because the fence protects the surveillance, which in turn protects the fence.

The purpose of the barbed wire/surveillance combination is to provoke a delay in response to a threat, and at

*In this section, Razac uses "virtual surveillance" to signify surveillance not visible to those controlled. He also uses "virtualization" to express rendering invisible the tools of surveillance and control. Translator's note.

Hungary, 1956. Hungarian troops remove fortifications on
the Hungarian-Austrian border, part of the attempt of the
liberal policy of Erno Gero to open up borders to the West.
Erich Lessing, photographer. (*Courtesy AKG Berlin*)

The place of death's absolute and arbitrary tyranny. Undated photograph. Dead soviet soldiers in a German prisoner of war camp, World War II. (*Archiv für Kunst und Geschichte, Berlin, courtesy of AKG*)

the same time to make a rapid and efficient defense, thanks to the information supplied by the surveillance. Thus, in its function, the barbed wire/surveillance is more temporal than spatial.[15]

Barbed wire appears to demonstrate that the modern problem of the political management of space cannot be resolved either by lightening barriers or by intensifying their ability to repel. Today, social control no longer relies on heavy separations such as fences; they are too visible, and they offer too many vulnerable points of attack. By progressing from physical enclosure to optical surveillance, the control of space has become discreet, and inter-active. It inverts the game of visibility. Whereas before one could make oneself hidden in order to attack a visible barrier, now it is the barrier itself that is hidden to the person who would attempt to breach it. Surprised, he is caught fully exposed to the reply that awaits him.

The invention of barbed wire was a step toward virtu-alizing spatial delimitation, because it favors the light over the imposing, speed over obstruction, transparency over opacity, the potential over the actual. But to virtual-ize control does not make it less real. It requires dispens-ing with fixed material devices and adopting others which react quickly and provide information on constantly changing situations. Instead of immobile towers and ramparts, modern power uses mobile tools which react only when necessary. This virtualization means more and not less control of space. Indeed, the strength and authority of power are increased.

Barbed wire can be considered a decisive stage in the history of virtualizing the political management of space. Yet, the symbol of power, which represents the capacity to enclose space, tends to weaken; it becomes a negative image of a brutal sovereignty, privileging the signs of domination over its efficacy as a tool. By the end of the Second World War, barbed wire began to be viewed as clumsy and outdated. More significantly, it had also become an almost universal symbol of oppression.

3

BARBED WIRE

TODAY

Barbed wire, although almost unchanged in the last hundred years, still marks a number of brutal social, political, and economic divisions. North American Indians continue to run up against barbed-wire fences. In Mexico, ever since the 1970s, barbed-wire fences surround fields and pastures, to the detriment of the Indian-style agriculture. "The differences of wealth between the ranchers and the Indian farmers are represented by barbed-wire fences. Wire enhances productivity, and so the fences deepen the gulf between the ranchers' and Indians' methods of production. One group has grown wealthier; the other has grown poorer and been forced to leave the land."[1] In Brazil, big landowners monopolize huge areas, often illegally, by using barbed wire and terror to get rid of Indian and mixed-race small farmers. "In the Parrot's Beak, a region where land is farmed collectively and in small settlements of twenty to thirty families, we have verified that shady and disreputable land brokers have surrounded communities with barbed wire. In one of these villages . . . the military police of Goiás burned down thirty-two houses in front of their owners."[2] Mexican Indians illegally trying to enter the United States encounter barbed wire at the border between the two countries. In the American Southwest, barbed wire has acquired such significance that in 1996 an Indian artist offered the University of New Mexico a sculpture representing three Aztecs migrating to the United States. Barbed wire crowns the top of the work. The artist explained that "everything in this sculpture is a symbol," and that barbed wire, found both in the sculpture and on the frontier, "is a dehumanizing part of our lives."[3]

There are several other barbed-wire frontiers. While Morocco was at war with the Polisario Front, the Moroccans built, between 1980 and 1987, a 1,500-mile wall in the western Sahara. It protected "160,000 Moroccan soldiers, who were dug in behind their minefields and barbed wire, as well as radar and other forms of electronic surveillance."[4] In Cyprus, the north, occupied by the Turkish army since 1974, and the Greek south are separated by 115 miles of barbed wire, the "last wall in Europe."[5] The Turks call this line "Attila." In southern Lebanon, the area occupied by the Israelis was separated from the rest of the country. In April 1999, when the Israeli army occupied the village of Arnoun, the Israelis "cut off the village from the rest of Lebanon, and thus annexed it de facto to the zone which it occupied. . . . They erected a sand embankment more than six and a half feet high at the village entrance. Barbed wire was installed around the village, and tractors dug a ditch along the road."[6]

In the same way, the Palestinian territories are separated from Israel. Gaza, a closed area, is surrounded by barbed wire. At the Erez checkpoint, "in order to enter Gaza, you must leave your car in a parking lot, walk a kilometer, show your passport three times. Then you walk alongside a tunnel of chain-link fence [plus barbed wire] which, from four in the morning, spits out Palestinians as if they were cattle, going to earn their 120 shekels a day in Israeli fields, construction sites and factories."[7]

Today, refugees trying to flee the war in Chechnya
come up against Russian checkpoints. At the end of
November 1999, at the frontier with Ingushetia, the line
of vehicles stretched out over almost two miles. "A group
of dirty and cold men and women stood permanently in
front of barbed wire rolled across the road. On a sign was
a handwritten message: 'Attention, frontier post. Beyond
this limit, you'll be shot at.' "[8]

Although wire fences have almost disappeared from
the frontiers of the European Community, immigrants to
the member states find a "new protective wall; one without barbed wire, minefields, guard towers and trenches,
but just as efficient and much more deadly."[9] Along the
Spanish coasts, "the police . . . do not shoot at them [ille-
gal immigrants crossing the Strait of Gibraltar]: they
gather them up in nets and take them back, dead or alive,
to their point of departure."[10] In Austria, "the barbed wire
which surrounded 'the Eastern Bloc' has been dismantled, replaced with a discreet but implacable form of surveillance to create the exterior frontier of the European
Community."[11] "Holding" camps have proliferated, making the phrase "a barbed-wired Europe" more appropriate than ever.[12] At Arenc, near Marseille, chain-link
fences and barbed wire surround the detention center

from which 1,492 people were expelled in 1998 (out of the total 1,752 who passed through it). A humanitarian organization has found that the camp's "conditions of confinement are inhumane . . . [and] unworthy of a democracy."[13] Belgium has six centers of this kind, whose explicit objective is to expel 15,000 people per year. "All these centers look and function like prisons: double lines of fences, often topped with barbed wire, surveillance cameras, guards, severely limited access [and particularly harsh rules]."[14] Italy, long accused of laxity concerning immigration, has also opened "detention" centers. "Barbed wire, police cordons, attempts at mass escape, manhunts: some Italians talk about the violation of civil liberties, calling the centers concentration camps."[15]

Side by side, the camps and barbed wire continue their uninterrupted march through history. A photograph shows Bosnian Muslims massed behind the barbed wire of a camp; the image is mounted next to a photograph of a Nazi camp. Historical continuity is represented by the barbed wire, which seems to extend from one photo to another, almost connecting them.

Refugee camps, where barbed wire is ubiquitous, often closely resemble concentration camps. Their barracks are "lined up in squares and are surrounded by chain-link fences and barbed wire."[16] In Palestine, Lebanon, Syria, and Jordan, hundreds of thousands of Palestinian refugees have been living in the camps for thirty to fifty years. Although the refugees' material situation has improved since the camps began, some still live in dilapidated prefabricated structures, and all remain exiles. "The camps reflect the triple marginalization—economic, social, and national—of which the refugees are the victims."[17]

Recently, in Kosovo, the refugee camps were surrounded by chain-link fences and barbed wire. At Kukës, in May 1999, four hundred Kosovars returned from captivity. "When their buses arrived at the edge of the camp, madness suddenly erupted. Hundreds of hands were stretched out behind the barbed wire; some inmates cried, some tore their faces on the barbed wire," in the hopes of recognizing a husband, a father, a son.[18]

The worldwide success of barbed wire cannot be denied. It is still associated with a brutal, authoritarian delimitation of space. It has, however, disappeared from our daily landscape, that is, from the urban zones of liberal democracies. Even refugee holding camps limit its use, and will no doubt soon abandon it completely. Nevertheless, barbed wire's image has not improved since the Second World War. Because it is a symbol of prison

Palestinian refugees behind barbed-wire fence, Gaza, 1989.
Peter Tumley, photographer. (*Courtesy of Corbis*)

and concentration camps and of fascist oppression, it still sums up the various barbarities and tyrannies with which it is associated. The logo of Amnesty International, the organization which protests imprisonment and torture for political reasons, is a burning candle surrounded by barbed wire.[19] The symbolism of barbed wire can change only when the wire is destroyed. In 1989, the leaders of Hungary made an overture toward the West. "As a symbolic act, [the minister of foreign affairs] . . . and his Austrian counterpart cut in two a piece of the barbed wire marking the Iron Curtain between Austria and Hungary."[20]

Any government using barbed wire risks alienating its people, and will use it only when the political costs do not exceed the advantages. Productivity depends on the efficient management of flux, and the ordering of space must ensure the greatest control possible over circulation, at the same time allowing for freedom of movement to the greatest possible extent. This is especially true for places of consumption, such as shopping malls and theme parks, which employ the most discreet tools of control—tools which are no less violent for those deemed undesireable.

The fence and the door are either open or shut, allowing or denying access. Barbed wire, by means of its necessary connection with surveillance, represented a refinement in the techniques of filtering, and an interactive enclosure, allowing for an adaptive response to events based on information furnished by surveillance. New optical-electronic technologies go further in this sense, by delimiting space but not barring access to it physically. These modern tools work directly on individuals, but indirectly on whole populations.[21]

The checkpoint is the archetypical barrier, which permits exits and entrances according to predefined criteria. Even in peacetime these checkpoints use barbed wire. In Brazil, private police often block off streets with barbed wire to restrict access to a commercial district or an affluent residence. But liberal democracies avoid displaying such signs of violent repression. They prefer to control passages unobtrusively and to allow maximum movement. To this end, democracies can avail themselves of several measures.

One of them is the security guard, whose role is not to repress but to mediate; he should dissuade, explain, supervise, and guide. In a shopping mall or theme park, he is there to gently discourage or politely expel anyone whose behavior excessively deviates from what the management considers the norm. When a guard ejects someone, he is rectifying an error, since undesirable persons should have been barred at the entrance. There the selection is invisible but firm. Without strong-arm tactics, it keeps out undesirables.[22] The guard is connected to two measures which give him information: electronic gates and video cameras.

The electronic gate detects the invisible: metal, magnetized objects, and, with X rays, the outlines of concealed and suspect objects. Electronic exits are used to spot stolen articles. All the customers undergo a thorough but invisible body search which does not disturb them. Only someone hiding a stolen article triggers both an alarm and the arrival on the scene of the security guards. In airports, courthouses, and athletic stadiums, X rays at electronic gates assure security by detecting potentially dangerous objects.

Camera surveillance systems are suspended over the heads of the people in all these places. Since 1998 in France, a million closed-circuit video surveillance systems have been at work, nearly 150,000 of them in public locations. The cameras observe behavior in streets, stores, and stadiums. They also watch the entrances and exits of parking lots, shops, amusement parks, and homes, all of which are more or less closed.

In real time, these systems can warn security guards when "at-risk" people enter, who may then be expelled or watched more closely. Thanks to the recording device, it is even possible to track "undesireables" after the fact, by noting on the tape those who broke the rules. In reviewing the tape, one might also establish a list of "shady" persons likely to cause trouble. The tape thus allows for the almost instantaneous correction of errors, recapturing those who should not have been allowed in in the first place.

Recent developments in video surveillance are aimed at speeding it up through automation. "Research is currently being carried out to develop special computer programs designed to automatically spot a sought-after person who has been recorded on videocassettes among a group of several other people. The perfection of surveillance programs will make possible an almost 'objective' recognition of behavioral abnormalities, atypical dress, or particular ethnic origins."[23]

Digitalizing optical signals means that they can be processed with a hitherto unknown speed. We go, then, from

a "passive optic" to an "active optic," which accelerates
and strengthens the ability of the surveillance tools to
react.[24] All of this creates the possibility of invisible and
automatic spatial delimitations. Working together, cameras and computers can choose whom to admit and
whom to refuse. A simple luminous signal indicates their
selections: green if yes, and red if no. One might possibly
argue with a guard, but one cannot argue with a computer. Furthermore, "the tendency of the market will be
to replace surveillance agents with electronic measures."[25]

Urban spaces are more and more divided into zones
with their own access requirements and behavioral criteria. Shopping centers separate those with money from
those without it. They promote behavior which incites
the solvent customers to spend. The objective of theme
parks is to "exploit the consumer-client's desire in a
space planned out to the millimeter while excluding
undesireables, and all of it under the guise of progres-
sive interests."[26]

The most notable development of recent years is the
proliferation of "secure" housing developments. These
are communities whose inhabitants choose to live on land
clearly separated from the urban fabric surrounding
them. In the United States, more than eight million peo-

ple live in closed developments which are guarded by private police who use video cameras, automatic entrances, and magnetic cards. Gated communities are now springing up in France. In the south, in Toulouse, Bordeaux, Montpellier, Aix-en-Provence, the security of such places figures in property developers' sales pitches. One enters these sheltered complexes "only by giving a password. . . . [Their] gates are opened by remote control . . . [and they] have a system of sophisticated intercoms, constant video surveillance, and vigilant guards who make nightly rounds." All these measures give the residents a sense of tranquillity. They are also set in the context of a warm and hospitable atmosphere which offers prospective buyers the "intimacy and happiness associated with small towns." Barbed wire obviously has no place in these communities, since its presence would destroy the conviviality of these ghettos of the rich. Moreover, security measures which are too visible can only partially succeed. A "wall or an enclosure can be counterproductive to security because it attracts attention."[27] It can also reduce sales, for prospective owners do not want to see the means of segregation which they nevertheless seek. New technologies will soon create invisible and efficient divisions. When such barriers are achieved, the last political and social

United States Marines transport a detainee behind layers
of fencing and razorwire at Camp X-Ray in Guantanamo
Bay, Cuba, February 6, 2002. (*Photo by Chris Hondros/
Getty Images*)

obstacles to segregating space will be lifted. So it seems that the violence of power is unacceptable only when we see it in action.

In the city, spatial hierarchy is produced by even more insidious openings and closings. The interfaces of urban access remove certain categories of the population and attract others by requiring a specific level of wealth, and by emitting symbolic signals with regard to social, ethnic, or cultural affiliation. "Space is one of the places where power affirms itself in the most subtle of forms, whether symbolic or invisible."[28] This discreet violence invites certain social or ethnic groups to occupy a given space and dissuades others from doing so. It does not forbid. This is done in such a way that choosing and rejecting seem to result simultaneously from the level of income and from a free choice based on personal preferences. The violence of the making of social hierarchies in space is thus partially blurred and camouflaged.

The conjunction of direct and indirect choices permits a rigid organization of social space, which assures that each person is automatically and discreetly relegated to a specific place. "[Thus] the most financially deprived are kept at a distance from the most socially coveted goods, symbolically and physically. . . . The lack of money intensifies social immobility: it ties a person to one spot."[29] Everyone carries his social/spatial status on his shoulders. The different signs of this status enable the interfaces of access to make selections appropriate to the requirements of a specific location. Status creates a social scale which determines who may go into areas considered attractive. Those who cannot enter any place are condemned to wander in a "social and spatial no-man's-land."[30] For them, there is only the exterior, the outside.[31] It is the dumping ground of democratic and liberal inclusion. This nonplace is the opposite of one ruled by biopolitics, which nurtures life. It is a discreet site where people are abandoned to die; perhaps one day it will become an equally unobtrusive place where people will be put to death.

NOTES

1. THREE HISTORICAL LANDMARKS

1. Henry D. McCallum and Frances T. McCallum, *The Wire That Fenced the West* (University of Oklahoma Press, 1965), pp. 31–32.

2. Pierre Mélandri, *Histoire des États-Unis depuis 1865* (Nathan Universite, 1984), p. 26.

3. James O. Adams, *Eleventh Annual Report of the Board of Agriculture for the Year 1881* (New Hampshire, 1882), pp. 31–32.

4. Walter P. Webb, cited in Claude Folhen, *La Vie quotidienne au Far-West (1860–1870)* (Hachette, coll. Littérature, 1974), p. 86.

5. Letter of Jefferson to William Henry Harrison, 27 February 1803, cited in Nelcya Delanoë, *L'Entaille Rouge. Des terres indiennes a la democratie americaine, 1776–1996* (Albin Michel, 1996), p. 54.

6. Roger Renaud, "On a jamais decouvert l'Amerique, on l'a niee," *De l'ethnocide,* 18 October 1972, pp. 12–13.

7. Cited in Janet A. McDonnell, *The Dispossession of the American Indian 1887–1934* (Indiana University Press, 1991), p. 6.

8. McCallum and McCallum, p. 204.

9. Cf. B. G. Trigger and W. E. Washburn, *The Cambridge History of the Native People of the Americas* (Cambridge University Press, 1996), vol. 1, part 2, p. 216.

10. Delanoë, p. 72.

11. Frederick Jackson Turner, *La Frontière dans l'histoire des États-Unis* (PUF, 1963), p. 178.

12. McCallum and McCallum, p. 11.

13. Harriet G. and J. Mauduy, *Géographies du western* (Nathan Universite, "Cinema et Image," 1989), p. 175.

14. Raymond Bellour, *Le western* (Gallimard, Tel, 1993), pp. 129–30.

15. Ibid., p. 137.

16. Quotations are taken from King Vidor's film *Man Without a Star,* 1955.

17. Maréchal de Folard, cited in De Clairac (Chevalier), *L'Ingénieur de campagne ou traité de la fortification passagère* (1749), pp. 93–94. (Work available in the library of the Museum of the Invalides.)

18. Cf. "foxholes," a term dating from the First World War, for holes in which soldiers protected themselves from enemy bombardments.

19. The French term for such an obstacle is *abbatis,* a word dating from the twelfth century.

20. A. Brialmont (Lieutenant General), *La Fortification du champ de bataille* (1878), p. 21. (Work available in the library of the Museum of the Invalides.)

21. Ibid., p. 331.

22. Réginald Kann, *Journal d'un correspondant de guerre en Extrême-Orient* (Calmann-Levy), pp. 265–66.

23. Soloviev (Captain), *Impressions d'un chef de campagne* (Librairie militaire, R. Chapelot et Cie, 1906), p. 34. (Work available in the library of the Museum of the Invalides.)

24. Pierre Miquel, *La Grande Guerre* (Fayard, 1983), p. 192. This report of improvization should be qualified. French army engineers perfectly mastered, at least in theory, the technique of trenches and secondary defenses. In August 1914, around the fortress of Maubeuge "1,500,000 stakes [were planted] around which thousands of miles of barbed wire intertwined; these immense networks, covering a total surface of 2,471 acres, soon surrounded each fortress." In Christian-Froge (director), *14–18. La Grande Guerre, vécue, racontée, illustrée par les combattants* (Librarie Aristide Quillet, 1992), p. 88.

25. Joseph Reinach, *La Guerre sur le front occidental. Étude stratégique. 1914–1915* (Bibliothèque Charpentier, 1916). (Work available in the library of the Museum of the Invalides.)

26. Miquel, p. 193.

27. Ministère de la Guerre, École de fortification de campagne, "Notice relative à l'installation des résaux de file de fer" (Imprimerie G. Delmas, 1914), pp. 25- 26.

28. Ibid., pp. 34, 43, 45; and Archives de Vincennes, Service historique de l'armée de terre, 16N855, 2V124.

29. Pierre Miquel, "L'annee 14," in *14–18: Mourir pour la Patrie,* a work published by the periodical *L'Histoire* (Le Seuil, 1992), p. 114.

30. Cf. Ernst Jünger, *Orages d'acier,* French translation, Christian

Bourgois, 1970 (Le Livre de Poche biblio., 1999), pp. 14, 70, 86, 193, 245, 273, 313.

31. Jacques Meyer, "La Vie quotidienne des sóldats pendant la Grande Guerre," in Miquel, *14–18: Mourir pour la Patrie,* p. 61.

32. A pair of metal or wooden Xs joined by bars and surrounded by barbed wire.

33. J.-P. Gueno and Y. Laplume, *Paroles de poilus, lettres de la Grande Guerre* (Tallandier, Historia, 1998), p. 13.

34. Jünger, p. 156.

35. Gueno and Laplume, p. 14.

36. Micheline Kessler-Claudet, *La Guerre de quatorze dans le roman occidental* (Nathan Universite, Collection 128, 1998), p. 44.

37. Ibid., pp. 34–49.

38. Henri Barbusse, *Le Feu. Journal d'une escouade* (Flammarion, 1965; Le Livre de Poche, 1997), p. 245.

39. Roland Dorgelès, *Les Croix de bois* (Albin Michel, 1919; Le Livre de Poche, 1992), p. 166.

40. Jack Sweeny, "Letter to a Girlfriend," 1916, cited in *The Spartacus Internet Encyclopedia*.

41. George Coppard, "With a Machine Gun to Cambrai," 1916, cited in *The Spartacus Internet Encyclopedia*.

42. "KZ" is the abbreviation for Konzentrationslager, used by the prisoners.

43. Stanislas Kozak, cited in Marcel Ruby, *Le Livre de la deportation* (Robert Laffont, 1995), p. 331.

44. Eugen Kogon, *L'État SS, le système des camps de concentration allemands* (Le Seuil, Points Politique, 1970), p. 52.

45. Ady Brille, *Les Techniciens de la mort* (FNDIRP, 1979), p. 163.

46. Hannah Arendt, French translation, *Les Origines du totali-*

tarisme, vol. 1, *Le système totalitaire* (Le Seuil, Points essais, 1972), p. 183.

47. These remarks are relevant to the Nazi form of the concentration camp. The Soviet camp was considerably different.

48. David Rousset, *Le Système concentrationnaire* (Hachette Littératures, Pluriel, 1998), p. 50.

49. Ruby, p. 379.

50. Robert Antelme, *L'Espèce humaine* (Gallimard, Tel, 1957), p. 34.

51. Here there is obviously no question of favoring one view over the other—the camp as a place of torture or as a totalitarian city. What is pertinent to our subject is the role of barbed wire in the concentration camp.

52. Primo Levi, *Si c'est un homme* (Juillard, Pocket, 1987), p. 44.

53. Ibid., p. 29.

54. Antelme, p. 21.

55. Rousset, pp. 103, 106.

56. Kogon, p. 192.

57. Abraham Bomba, in Claude Lanzmann, *Shoah* (Gallimard, Folio, 1985), p. 162.

58. Levi, p. 181.

59. "Les Barbelés du bois de Boulogne," *Le Monde,* 7 August 1946.

60. François Cochet, *Les Exclus de la victoire. Histoire des prisonniers de guerre* (1992), p. 232.

61. Kogon, op. cit.; Milton Meltzer, *Never to Forget: The Jews of the Holocaust* (Dell, 1976).

62. Claude Laharie, *Le Camp du GURS* (1985); Gilbert Badia *Les Barbelés de l'exil)* (PUG, 1979); Feuchtwanger, *Le Diable en France* (Godefroy, 1985); Sylvain Kaufmann, *Au-delà de l'enfer* (Séguier,

1987); Levi, op. cit.; Anne Grynberg, *Les Camps de la honte* (La découverte, 1991); the trimonthly bulletin of the Auschwitz Foundation.

63. Andrej J. Kaminski, *Konzentrationlager 1986 bis heute*. See also Léon Poliakov, *Auschwitz* (collection Archives), 1964.

64. Joë Nordmann and Anne Brunel, *Aux vents de l'histoire, mémoires* (Actes Sud, 1996), p. 185.

2. BARBED WIRE AND THE POLITICAL MANAGEMENT OF SPACE

1. Once again, barbed wire on its own did not effect results. It was a tool which in fulfilling certain needs or desires created consequences.

2. Cf. Alexis de Tocqueville, *De la Démocratie en Amérique* (GF-Flammarion, 1981), vol. 1.

3. Alain Brossat, *L'Épreuve du désastre. Le XXieme siècle et les camps* (Albin Michel, Idées, 1996), p. 28.

4. Michel Foucault, *La Volonté de savoir* (Gaillimard, Tel, 1976), p. 179.

5. Henry D. McCallum and Frances T. McCallum, *The Wire That Fenced the West* (University of Oklahoma Press, 1965), p. 138.

6. Georgio Agamben, *Homo Sacer* (Le Seuil, L'ordre philosophique, 1997), p. 132.

7. Ernst Jünger, *Orages d'acier,* French translation, Christian Bourgois, 1970 (Le Livre de Poche biblio., 1999), pp. 275–287.

8. James O. Adams, *Eleventh Annual Report of the Board of Agriculture for the Year 1881* (New Hampshire, 1882), p. 32.

9. Bernard Dort in Raymond Bellour, *Le western* (Gallimard, Tel,

1993), p. 60.

10. Henri Barbusse, *Le Feu. Journal d'une escouade* (Flammarion, 1965; Le Livre de Poche, 1997), p. 55.

11. Extract from a speech of Adolf Hitler, cited in Charles Dubost, *Les Procès de Nuremberg, l'accusation française,* vol. 5 of *La Politique allemande d'extermination* (Tristan Mage, 1992), p. 8.

12. Gordon J. Horwitz, *Mathausen, ville d'Autriche. 1938–1945* (Le Seuil, Coll. "Libre examen," 1992), p. 196.

13. Cf. Jean-Louis Cohen, " 'La Mort est mon projet': architecture des camps" in Francois Bedarida and Laurent Gervereau (editors), *La deportation, le systeme concentrationnaire nazi* (La Découverte, Les publications de la BDIC, 1995), pp. 35–36. Contrary to Jean-Louis Cohen, the architecture of the camp does not seem to us panoptic. The discussion could have bearing on the panoptic character of the surveillance of the kapos or the effect of terror.

14. On the Panopticon, cf. Jeremy Bentham, *Le Panoptique,* French translation, Pierre Belfond (1977). On its diffusion as a general disciplinary principle, cf. Michel Foucault, *Surveiller et punir* (Gallimard, Tel., 1975), pp. 228–264.

15. The effect is spatial but the functioning is mainly temporal.

3. BARBED WIRE TODAY

1. Hugues Cochet, *Des barbelés dans la sierra* (ORSTOM, 1993), p. 287.

2. "La terre et le sang," extracts from *Dial,* nos. 983 and 1002, in *Le Monde Diplomatique,* April 1985.

3. Cited in Roberto Rodriguez, "Barbed wired for controversy," www.indians.org

4. Jean Ziegler, "Quinze ans de conflit au Sahara Occidental," *Le Monde Diplomatique,* March 1989. See also Mariano Aguirre, "Vers la fin du conflit au Sahara Occidental," *Le Monde Diplomatique,* November 1997, on the resolution of the conflict, the end of fighting in 1991, and the referendum which was scheduled but has constantly been canceled.

5. Niels Kadritzke, "Chypre, otage de l'affrontement entre Athènes et Ankara," *Le Monde Diplomatique,* September 1998, and Eric Rouleau, "La partition s'enracine a Chypre," *Le Monde Diplomatique,* October 1996.

6. "Un village libanais une nouvelle fois annexé a la zone occupée par Israel," *Le Monde,* 17 April 1999.

7. Béatrice Guelpa, "Gaza, écran total," *L'Hebdo,* 30 December 1998.

8. Anne Nivat, "Les réfugiés liés au bon vouloir russe pour passer la frontière," *Liberation,* 30 Nov. 1999.

9. Juan Goytisolo, "Un nouveau mur de la honte," *Le Monde Diplomatique,* October 1992.

10. Ibid.

11. Lorraine Millot, "L'Autriche frissonne sous les vent d'Est," *Saturday and Sunday,* 13 June 1999.

12. *A Barbed-Wire Europe* is also the title of a collection of texts presented by Jean-Pierre Perrin-Martin, published by L'Harmattan, Paris, 1998. Perrin-Martin is also the director of "Barbed Wired Europe" and coauthor of *La Rétention* (L'Harmmatan, 1996).

13. Cf. Pedro Lima and Régis Sauder, "Arenc, inhumaine antichambre du départ," *Le Monde Diplomatique,* November 1999.

14. Cf. Laurence Vanpaeschen, "En Belgique, un arsenal répressif

contre les étrangers," *Le Monde Diplomatique,* January 1999.

15. Aloise Salvatore, "L'Italie ne veut plus être un 'aubaine' pour les irréguliers," *Le Monde,* 20 August 1998.

16. Frédéric Bobin, "Les dernières heures des 'boat-people' vietnamienes à Hongkong," *Le Monde,* 18 April 1995.

17. Nadine Picaudou, "Dispersion, résistance et espoirs des exiles palestiniens," *Le Monde Diplomatique,* July 1992.

18. Florence Aubenas, "Prisonniers et boucliers des Serbes," *Libération,* 24 May 1999.

19. See also Egon Larsen, *Une flamme derrière les barbelés, histoire d'Amnesty International,* 1979. Also the Internet site of Amnesty Canada features a brief animated cartoon showing a piece of barbed wire whose barbs gradually are transformed into a flight of birds.

20. "En aucun cas, nous n'aurions recours aux vielles méthodes," interview with Gÿula Horn, former Hungarian minister of foreign affairs, in *Le Monde,* 5 November 1999.

21. The term "interface" is interesting because of its generality. It designates any measure or apparatus which regulates the passages between two distinct systems. Interface is simultaneously the surface of contact between two zones (faces) and the procedure which regulates the exchanges between these two zones (inter). "Access," "threshold," and "frontier" are too passive because they only designate the limit, not the action of selection.

22. The procedure consists of detecting potential risks to the optimal functioning of the place in question. Such risks are appearance (foreign ethnicity, "shady people," lack of visible financial means, incorrect dress) and behavior ("inappropriate" nonchalance, brusque gestures, a loud voice, aggressiveness).

23. André Vitalis, "Être vu sans jamais voir. Le regard omniprésent de la vidéosurveillance," *Le Monde Diplomatique,* March 1998.

24. On the difference between the passive direct optic and the active indirect optic, see Paul Virilio, *L'Inertie polaire* (Christian Bourgois, coll. "choix-essais," 1994).

25. Alexandre Garcia, "La vidéosurveillance se généralise dans les lieux publics et les entreprises," *Le Monde,* 6 August 1998.

26. Susan G. Davis, "Quand les parcs à thème gangrènent les villes," *Le Monde Diplomatique,* January 1998.

27. J.-P. Besset and Pascale Kremer, "Le nouvel attrait pour les résidences sécurisées," *Le Monde,* 15 May 1999.

28. Pierre Bourdieu, "Effets de lieu," in *La Misère du Monde* (Le Seuil, Libre Examen, 1993), p. 163.

29. Ibid., p. 165.

30. Robert Castel, *Les Métamophoses de la question sociale, une chronique du salariat* (Gallimard, Folio essais, 1995), p. 666.

31. This "exterior" is not necessarily the street, but any place which, for only certain people, is a place of abandonment. This place is temporary and accompanies the undesirable wherever he goes.

INDEX